Quantum Resonance Frequency

A Theory on the Fundamental Nature

Of Energy and Matter

In the Universe

By Sean Allen Ellis

2015

I wish we could derive the rest of the phenomena of nature by the same kind of reasoning

From mechanical principals; for I am induced by many reasons to suspect that they may

All depend upon certain forces by which the particles of bodies, by some cause hitherto

unknown, are either mutually impelled towards each other, and cohere in regular figures, or are

repelled and recede from each other; which forces being unknown, philosophers have hitherto

attempted the search of nature in vain; but I hope the principals here laid down will afford some

light either to this or some truer method of philosophy.

Sir Isaac Newton

Quantum Resonance Frequency-

Quantum Resonance Frequency is a theory about the fundamental structure of matter in our universe that simply states all matter is pure energy bound together by the cohesive properties of the attraction between a negative and a positive charge creating a frequency that gives off an electrical charge and by doing so produces the fundamental properties of mass and the building blocks for the atomic structure. The world around us is filled with matter that we perceive to have a solid effect whether the matter is a solid, liquid, gas, plasma, or even photons and dark matter. The reality we perceive as human beings is due to the observations that we make from inside of the system that we are attempting to observe. All matter from the largest string of molecules a giant crystal lattice of multiple chemical compounds that we call dirt and rocks down to the tiniest atom and its sub components of quantum particles is just elements of pure energy acting and reacting to all of the other elements of pure energy in the universe around us. Large objects have a very large electrical charge because of the sheer volume of atoms that they are made of and therefore from our point of view, being built of the same elements, with a considerably sized electrical charge ourselves have the very strong appearance of what we refer to as solid matter. Subatomic particles on the other hand are just simple charges with fewer numbers and combinations of frequencies that appear more and more wavelike resembling the pure energy that all matter is formed out of more and more as it is broken down until all that is left is raw unbound energy flowing out.

Chapter 1

Basic Atomic Structure-

The atom as a property of matter is unique and even though an atom is built of several smaller sub atomic particles it is still the atom that every student being taught about the nature of the universe will begin their studies by learning about. The atom is a combination of protons, neutrons, and electrons usually possessing a symmetrically balanced electrical charge with the exception of hydrogen which is just a single electron orbiting a proton without a neutron. Protons, neutrons, and electrons can be observed moving through changes in their elemental charges or basic makeup but these charges are generally not stable and result in radioactivity, decay, or the formation of antimatter. The atom is the stable and balanced form of the pure energy that the universe around us has flowing through all of its dimensional perspectives. This makes an atom the ideal building block for the reality that we as observers exist inside of. The atom with its basic charge basic mass and the ability to combine and inter connect with other atoms in an infinite number of chemical combinations and possibilities is the reason that life itself can exist in unique individual aspects while still possessing the fundamental traits of the pure energy that has existed since the birth of the universe. The universe of pure energy given shape and form through the basic structure of the atom and its unique properties depending on the combination of electrical charges gives each atom their own characteristics and properties.

When mankind first started trying to make sense of the world we live in and understanding what were the basic building blocks of our reality humans believed that everything including atoms was made out of the basic elements earth, air, water, and fire. Through hundreds of years of dedicated human observation and scientific endeavors the vast majority of the human race has ruled out magic and the basic elements of earth, air, water, and fire as the basic building blocks and settled on the atom and the subatomic particles that the atom holds together. Greek philosophers first used the word Atomos to describe the smallest particles that all known matter could be cut down into. The concept was as easy as taking a piece of wood and cutting it repeatedly until finally you reached a piece of wood so small that it could never be reduced any further. We have found out through the use of nuclear weapons and particle colliders that Greek philosophers were wrong and you can cut, divide, and smash all matter down until there is nothing but massive amounts of energy left but still the name atom has been used for so long at this point that no one seemed to be bothered or took serious interest in petitioning that the name be changed. Even though we have advanced our understanding of the physical properties that make up the world around us including the chemical structure of the fibrous cells that make up wood and to most plant life we still use the term atom to express and to teach the basic building blocks of matter and energy.

An atom is made up of subatomic particles called electrons, protons, and neutrons. The atom can come in different combinations of these three subatomic particles and the different quantity of combinations of these subatomic particles will decide the structure and nature of the atom. A single electron orbiting a single proton is the most basic and most prolific atom throughout our universe and fuel the vast majority of stars are using for fusion, hydrogen. Two electrons orbiting two protons and two neutrons is a stable gas that is also a fuel stars commonly use for fusion, helium.

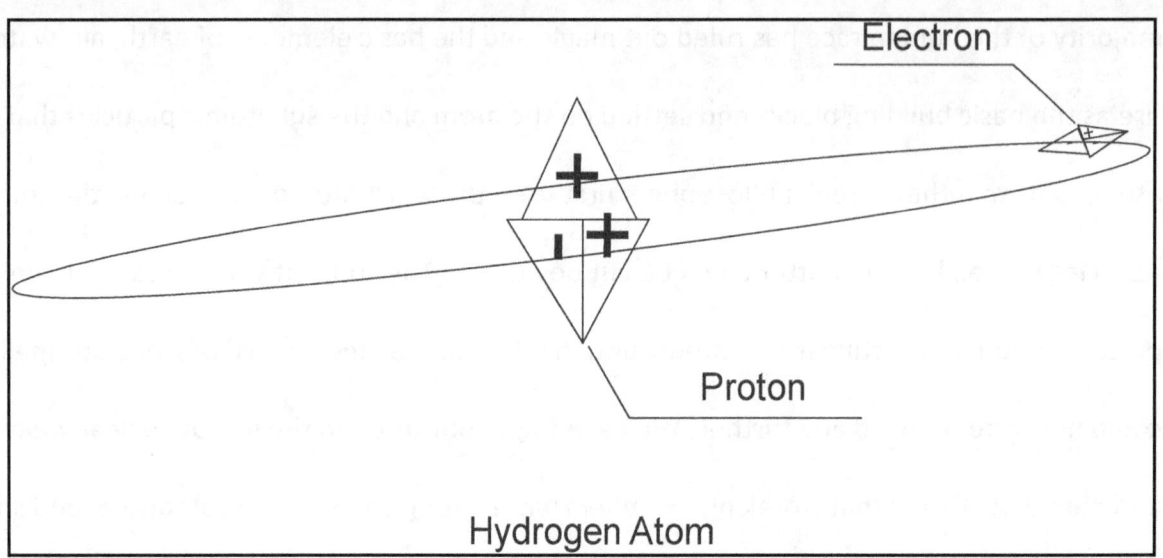

Figure 1- the hydrogen atom.

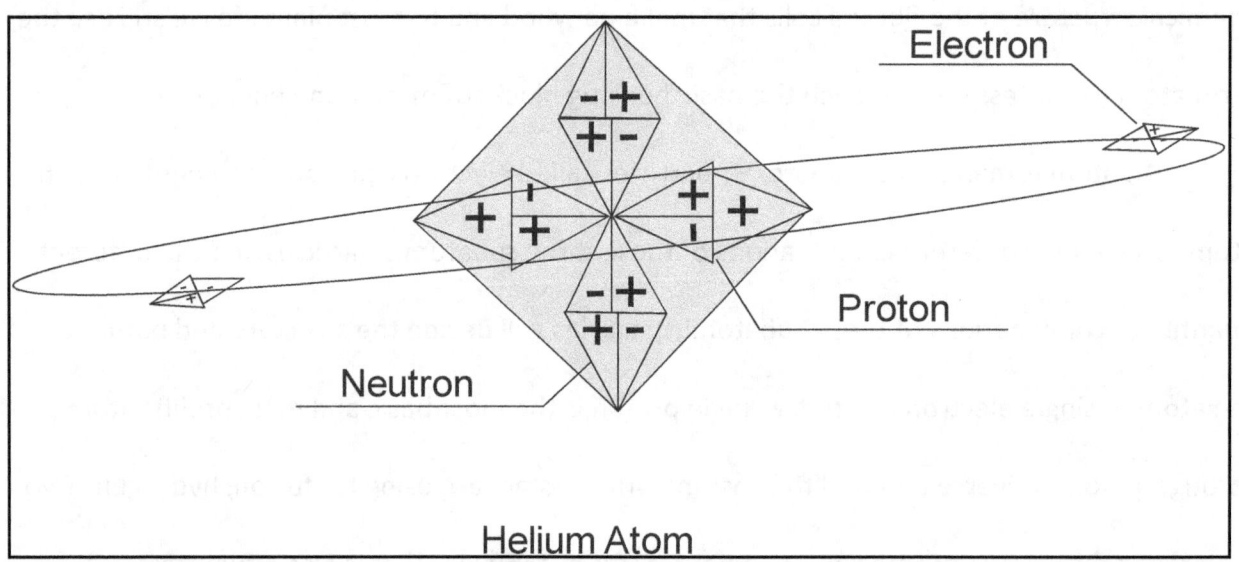

Figure 2- the helium atom.

The electron is the smallest and most active of these subatomic particles and possesses the most wavelike properties of all three. The electron has a negative charge and is the source of all electricity that we as humans use every day because of the electrons location in the outer orbital shells of all atoms. The nucleus or center of an atom is where the proton and neutron are constantly balancing the energy levels and therefore the frequency of the atom. Sometimes the proton and neutron keep the frequency of the atom balanced by telling the electron exactly what orbital shell or specific distance from the nucleus it has to orbit. The proton has a positive charge or possesses a frequency made up of both negative and positive energy with a slightly larger positive charge. The neutron possesses a neutral charge or has a frequency that is capable of maintaining a perfect balance between the positive and negative energy that this particle represents. These three subatomic particles that are used to build all atoms remain firmly bound together due to the properties of the pure energy that they represent and in doing so create a frequency that can adapt or modulate at different energy levels and still remain as a unique package of energy bound together to create an electrical charge that gives these subatomic particles the mass and properties of matter that we observe from inside of the universe.

When not bound to an atom; protons, neutrons, and electrons can produce large amounts of energy in the form of radiation or electricity. During radioactive decay protons and neutrons can be forcibly ejected from the nucleus of an atom producing large amounts of radiation that is harmful to life. Electrons can also be ejected from an atom as radiation but most human interaction with electrons is in the form of electricity. When severely excited or agitated theses subatomic particles begin to exhibit the traits of their more dangerous wavelike properties, pure energy. This blast of pure energy can damage the structure and nature of atoms that cannot withstand the energy being poured into them. Since all matter is a frequency of pure energy bound by its positive and negative forces this can be viewed like sound waves. Most of the time a human can sit and listen to music and pleasantly enjoy the soothing harmony of rhythms but if you pour too much energy into the sound system both the sound system and your ears can be damaged by the frequencies that are produced. When most humans think about frequencies they usually refer to sound waves or sometimes radio waves but the basic structure of an atom is a frequency. This frequency comes from the energy levels that make up protons, neutrons, and electrons. These frequencies are the expressions of the energy that is bound and given form by both the negative and positive forces being balanced charges of pure energy held together in tiny packets possessing positive negative and neutral charges. Theses subatomic particles each possess unique properties of mass and electrical charge and when in their most energetic states can damage the structure of other atoms that they come into contact with.

The basic simplicity of an atom is the number of different elements that can be created by different combinations of protons, neutrons, and electrons. A single positive charge being orbited by a tiny negative charge is a volatile gas that is the fuel in the heart of fusion cores of every star in the night sky. Eight positive charges balanced by eight negative charges give us oxygen in the very air we breathe. Since life exists inside of the system that we are observing our perspective is often clouded by what we believe our reality to be. It is hard to imagine everything around us as just being billions and billions of tiny electrical charges because we have a tendency to associate electrical charges with electricity and most of the objects in the world around us at least look fairly inert. We are not being electrocuted by everything we touch and interact with so we view it as something other than distinct combinations of electrical charges. The reality is since we are made up of the same substance we do not feel the electrical charges of the atoms around us. The experience that we receive is one of a solid environment due to the interaction of the charges of our own bodies. The atoms of our bodies are all chemically bound together possessing a very unique electrical quantity. The table in front of us is also made up of atoms that are all chemically bound together possessing their unique electrical quality. What we end up with instead of an electrical shock is a very solid sensation because it is extremely difficult to force these two similar electrical qualities into each other. Both electrical charges will act to prevent the other from connecting or attaching without energy being used to cause a different chemical reaction. The result is that to us the table is very solid even though it is just pure atomic energy chemically bound together in millions of layers. We are just as solid to the table as the table is to us because we all possess our own unique electrical charge that does not readily accept an outside quantity. All of this mass of electrical charges giving everything we know its own unique characteristics comes from a combination of electrons, protons, and neutrons forming different atoms that can then be chemically joined to larger and larger quantities of atoms to create every known molecule and structure that we see inside this system of living energy that is our universe.

The basic structure of an atom is also accompanied by its basic energy levels. The basic layout of an atom has the protons and neutrons in the nucleus arranged so that the positive and negative aspects of each frequency is ordered and lined up in positions that keep the energy fluctuations of the atom in the most stable layout that best promotes energy transfers between the nucleus and the orbiting electrons. Electrons can both orbit and remain stationary outside of the nucleus. The electron should remain in a constant state of motion representing normal kinetic energy constantly orbiting the nucleus for a perfect balance of energy levels and frequency. When an electron is chemically bound to another atom it will remain in a localized position vibrating and moving in small circles on the orbital shell. When electrons are chemically bound the nucleus of the atom should show signs of rotation to express the innate kinetic energy of an atom. The kinetic energy of an atom can increase or decrease depending on energy supplied from an outside source. This is usually represented as heat inside of most atoms where an outside energy source adds to the existing kinetic energy of the atom. The atom can also express this excess energy by moving electrons from one orbital shell, or energy level, to another expanding and contracting the overall size of the atom. This can also be expressed as orbital shells move outwards with higher energy levels of the electrons orbiting in these shells. The movement of an electron from one shell to another is referred to as a quantum jump. The electron jumps to a shell with a higher energy level to maintain the balanced frequency of the atom. The method that the atom is employing to move electrons from shell to shell is known as an Einstein-Rosen bridge or a wormhole. The nucleus of an atom is capable of producing and connecting a wormhole from an electron in an orbital shell to an orbital shell with a higher energy level in order to maintain the physical structure of the atom. The electrons position maintains the balanced frequency and energy levels of the nucleus and the potential capacitance that exists between the nucleus of an atom and the electron shells. This potential that exists between atoms and electrons that have held the frequency of the atoms that they have been attached to is the reason that quantum entanglement exists.

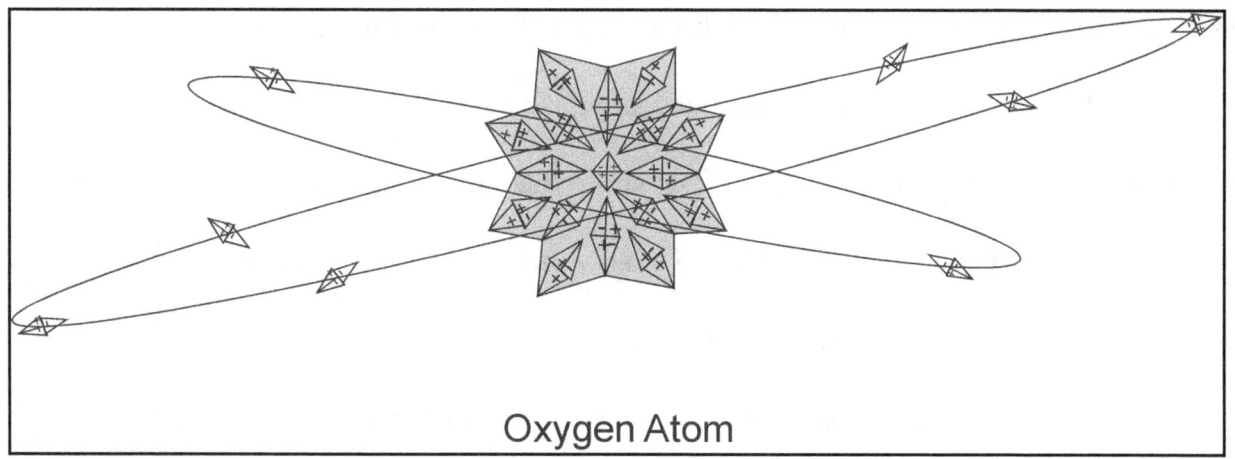

Oxygen Atom

Figure 3 Oxygen atom at normal energy levels.

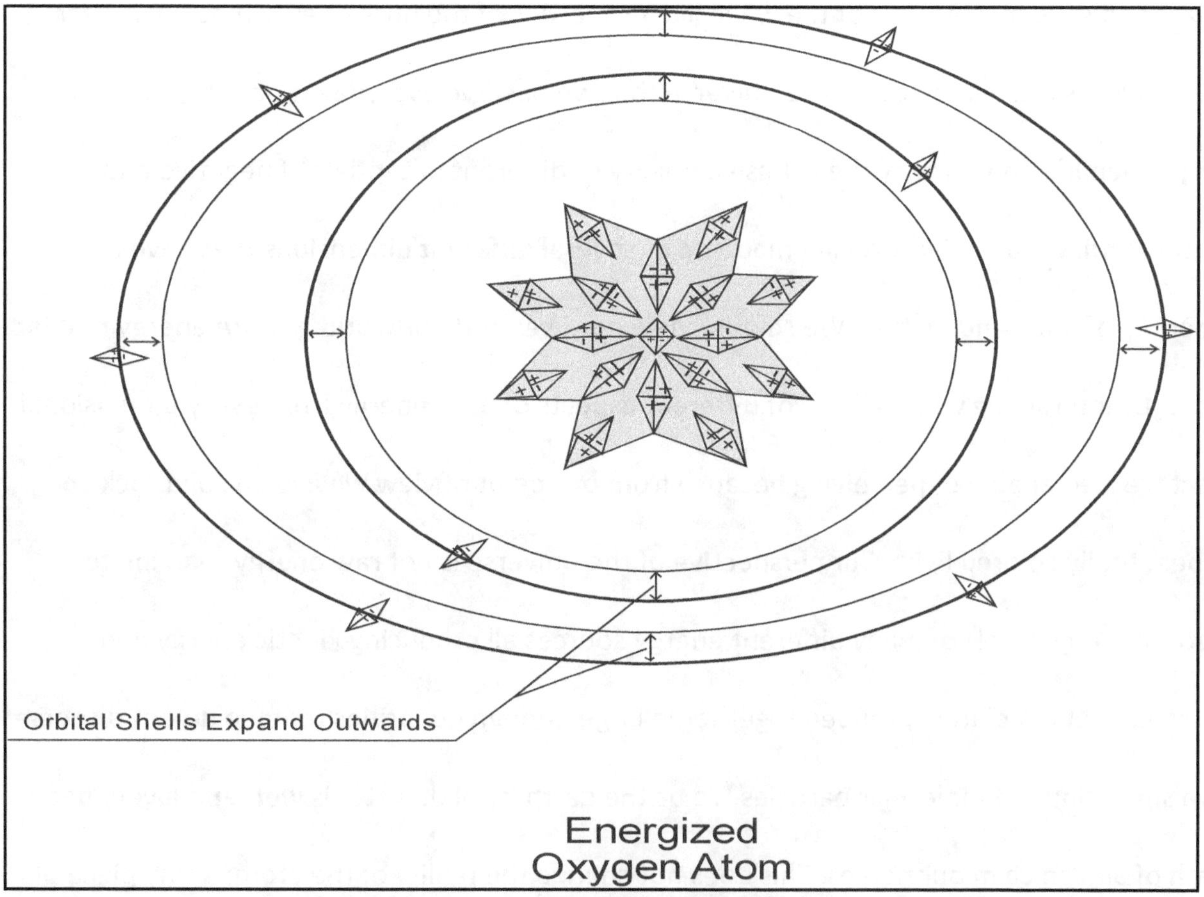

Orbital Shells Expand Outwards

Energized
Oxygen Atom

Figure 4- When an atom absorbs excess kinetic energy the orbital shells can expand outwards. The protons will use the positions of the neutrons to move themselves slightly outwards. This forces the electron shells to expand outwards so that the electrons can more efficiently disperse excess kinetic energy.

The fundamental fabric of wormholes will hold the potential frequency of the nucleus until the electron has become a stable part of another atom thereby changing its relativity and frequency to the new host atom. This fundamental stabilization is how an atoms basic physical structure matches its ability to keep both the energy that it is made out of and the kinetic energy from an outside source constantly balanced.

The basic structure of an atom is so much more than just three different subatomic particles that decided to hang out together. The subatomic particles of an atom; electrons, protons, and neutrons are the smallest subatomic particles that sill display more characteristics of mass and solid matter than of the wave like properties of the pure energy that these basic building blocks are made out of. The universe that we observe everyday is nothing but pure energy given life and form by the cohesive quality of different potentials of negative and positives until we perceive a reality made up of several different dimensions that have all the properties of mass and matter. We seldom view our lives and our world as pure energy that is constantly interacting with millions of different aspects of pure energy from every dimensional aspect we are capable of perceiving because from our point of view we live on solid rock and see beautifully colored light. Our perspective of the universe is not raw energy reacting to different energy levels of many different energy sources all producing kinetic energy and electromagnetic radiation grouped together in large enough quantities that we do not disperse into a small cloud of molecular particles. To us the earth is solid, water is wet, and a very deep breath of air can calm our nerves. This does not change the reality of the atom; a tiny piece of pure energy with its own unique electrical quality and charge bound together by aspects of positive and negative forces producing a stable frequency that does exist in a state where everything in its reality is another form of energy interacting and reacting to its presence and internal kinetic energy.

1.2 The Proton-

The proton is the positive charge or potential of the nucleus of all atoms. The proton inside of an atom is responsible for maintaining the positive aspects of the fundamentally balanced electrical charges of an atom. In every atom except hydrogen the proton maintains this balance by remaining relative in both position and energy levels to the neutron. From an electrical point of view this is very similar to using the neutron as a ground but at the subatomic level the proton is not emitting a charge that is cancelled out by the neutron. It is the combination of energy levels of the proton and electron that are balanced against the neutron and not cancelled out by it. The proton and the neutron match each other in energy levels, vibrations, spins, and in their positions inside of the nucleus of the atom. In a hydrogen atom without a neutron this is expressed by the gasses volatility and the natural inclination of hydrogen to bond to any element that has neutrons in their cores. Hydrogen in its natural state is found in the depths of space where the extremely low temperatures reduce the kinetic energy of both the proton and the electron. The proton is a solid particle that is made up of several pieces of the frequency of energy that we define as having mass and being solid matter. This means that the proton possess both a negative charge and a positive charge but has a larger positive potential due to the frequency of the energy that it fundamentally represents. Any subatomic particle constituents of the proton represents more wavelike properties than they do solids because when highly active the frequency of this pure energy still possess some of the charge, expressed as vibration, that gives matter mass and substance. Although as pure energy these properties disappear almost as quick as they are noticed as the energy loses the frequency that we would associate with that of solid matter. The proton as pure energy, the wave or frequency given properties of matter by the solidly bound forces of negative and positive, works in conjunction with the neutron to control the energy levels of the atom. One way that the proton will help to maintain energy levels is by simple expansion and contraction. The other way that the proton maintains energy levels inside of an atom is with quantum entanglement using the un-energized potential of a wormhole to remain relative in position and energy levels with the electron that it is paired with to complete the frequency of its own positive to negative charge that is the fundamental structure and strength residing inside of the atomic structure. The proton works in harmony with the electron to maintain a neutral state of existence even though it has a predominately positive charge while remaining relative to the neutron that becomes its basis for a balanced frequency.

The protons inside of an atom are responsible for producing and maintaining a positive charge. This is one of the reasons why pure energy can possess the aspects that we refer to as mass and solid matter. If all of the energy pouring through the universe possessed the exact same potential of charge it would not be capable of producing the effects we refer to as solid matter. For solid matter to exist both a positive and a negative charge have to become bound together. Even an extremely small electrical charge will become powerfully bound to the opposite charge and the energy levels inside of an atom are anything but small. Without the positive and negative aspects of energy giving tiny particles mass by representing the combined charges as a single vibrating frequency possessing the qualities of mass all energy would be neutral and flowing throughout the universe without form or structure or even variances' in energy levels from one end of the universe to the other. It would all just be neutral and without form. The proton takes both aspects of this energy and binds it giving it mass and a frequency based on its vibrational state from the aspects of possessing both negative and positive charges while maintaining a slightly higher positive potential and electrical charge.

The proton uses its position and energy levels inside of the nucleus to maintain the structural integrity of an atom. Everything in the universe is pure energy given form and structure through the strength of the atomic structure. All of this structured energy that we refer to as mass and matter is constantly moving. Everything that we know and see is constantly moving and interacting with everything else bouncing, vibrating, and constantly colliding producing what we refer to as kinetic energy. A proton uses its position in the nucleus to help balance out all of the kinetic energy that it is subjected to. As energy levels build the proton will match the electron in energy levels and a position similar to the electrons position in the orbital shells of an atom. The strongest structure that the atom can represent is the connective force of the positive and negative charge. The proton expends less energy to maintain this bond than it takes to overcome the strength the bond represents. The protons movement either by expanding and contracting or adjusting its position relative to the neutron creates a type of capacitance that helps the proton maintain its basic frequency. This capacitance is held between the proton and the neutron due to their energy levels and their distance from each other. This basic frequency is the structural balance of pure energy with a slightly higher positive potential than a negative potential. By matching the electron the proton is capable of balancing out the electrical charge of an atom. This creates a frequency that will perfectly match the frequency of the neutron inside of an atom. This means that even at the highest energy levels that an atom can be subjected to the cohesive strength does not diminish if anything the atomic strength should increase. This is what allows for the liquid and plasma states of most forms of matter without causing a chemical or fissionable change in the basic structure of the atom. The proton maintains the positive side of the cohesive structural strength of an atom by maintaining a position and energy level relative to that of the electrons absorbing most of the kinetic energy that the atom is exposed to and using the negative aspect of the electron to produce an energy level that is stable and relative to the neutrons in the nucleus of the atom.

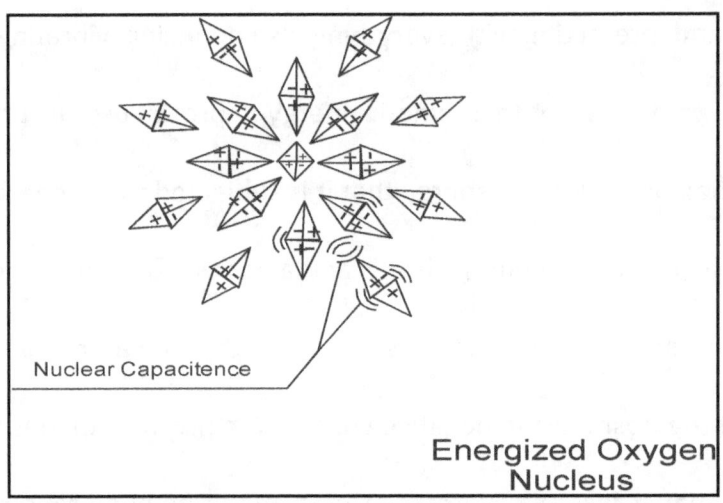

Figure 5- A type of nuclear capacitance can build up between the proton and neutron. This capacitance comes from the vibrational energy of the proton that has a quantum entanglement attachment to the electron when absorbing kinetic energy from outside of the atom.

In a hydrogen atom the proton is in a nucleus that does not have a neutron to help it balance out the electrical charge given to the atom by the kinetic energy of the universe around it. Hydrogen is a gas that is known to express the kinetic energy of the universe by constantly moving and expanding outward but since hydrogen does not have a neutron to give it a balanced perspective of a neutral charge it has a natural disposition to chemically bind itself to other atoms that do have neutrons expressing a balance of positive and negative charges. For hydrogen it is not just the need to have a complete orbital shell of electrons or hydrogen atoms would just bind to other hydrogen atoms. In the depths of space where the temperature lowers the natural kinetic energy the atom is exposed to hydrogen can be found as a gas because there is not as much of a buildup of potential kinetic energy. Hydrogen can be found as a gas in clouds that stretch trillions and trillions of miles across the galaxy. Here on earth the temperature causes hydrogen to constantly be in a higher energy state. The proton and electron still maintain a constant balance between the positive and negative aspects of energy but without a neutron to anchor the proton and give it a relative balanced perspective hydrogen is constantly combining with almost any atom that it comes into contact with. Unlike almost every other known element at temperatures found here on earth hydrogen will not be found as a natural gas it will only be found in chemical combinations with other elements. Hydrogen gas does not have a neutron in the nucleus for the proton to use as a neutral perspective therefore in temperatures found here on earth hydrogen will combine with other elements to help protect its atomic structure from the kinetic energy that it is subjected to.

The proton is pure energy represented as solid matter by the energized state that comes from its slightly higher positive charge bound to a negative charge giving the proton less wavelike characteristics of pure energy than the smaller pieces of its subatomic makeup. This simply means that unlike smaller subatomic particles the proton has more mass because it has a large electrical potential difference between the positive and negative aspects of the energy that it is made out of. This gives the proton a greater sense of what we refer to as solid matter where the smaller subatomic particles exhibit signs of being less like matter and more like the wave that they become due to their loss of potential difference between charges. The wavelike characteristics of the constituent building blocks of a proton can be described as the breakdown of the cohesive forces that give energy the properties of mass and matter. These smaller subatomic particles exist as pieces of the potential difference that holds the protons together for very short energetic periods of time. These expressions of energy are waves because they do not possess the strength of the atomic structure that comes from the strong binding of a positive and negative charge. Unlike other subatomic particles the proton has strong electrical qualities that give it mass and the properties of matter leaving the proton neutron and electron with considerably less wavelike characteristics than other subatomic particles.

The proton is capable of using kinetic energy through expansion and contraction. Just like solid matter that we see and interact with on a daily basis the proton uses energy that is registered as heat to expand matching the shells of electrons as they expand outwards. The proton does not change sizes like a hot surface does in sunlight. The proton actually moves outwards from its position in relation to the rest of the nucleus. This causes the atom to become slightly larger. The proton is absorbing some of the kinetic energy that the electron is dispersing this causes the proton to move closer to the neutral state or charge. The proton and neutron have mass due to the vibration of the bound potential energy that they are fundamentally built upon. For electrical charges the proton is slightly more positive than the neutron but since the neutron is neutral there is no attraction of positive to negative. Both the neutron and proton have a similar electrical quality because of the pure energy that causes the minute vibrations in matter that we define as mass. This similar electrical quality has as much of a repelling effect as the positive to negative electrical charge has as an attracting force. This keeps the atom perfectly balanced and bound together. During expansion this causes the atom to become slightly larger without the proton, neutron, or electron actually increasing in size.

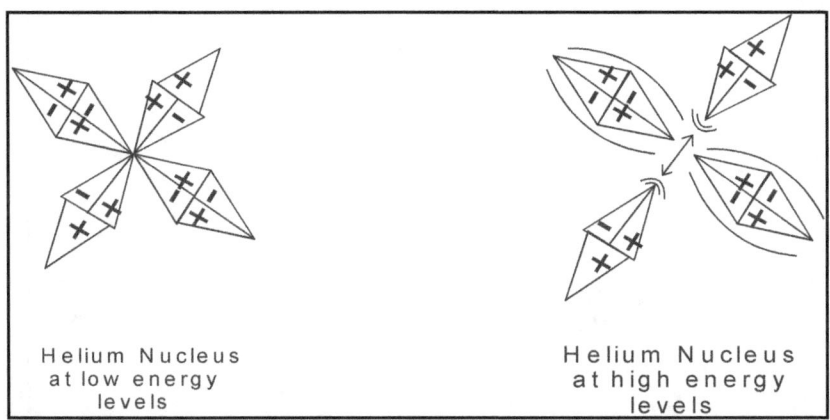

Figure 6- the nucleus of an atom expands to correspond to the electrons orbital shells expansion. Because of quantum entanglement the proton absorbs some of the kinetic energy that the electron is dispersing. This gives it a more neutral charge causing it to slightly push away from the neutron in the nucleus.

This higher state of kinetic energy that causes expansion is one of the ways in which a proton helps to maintain the balance between negative and positive charges holding the atom together. The atom is absorbing energy and this causes the electrons to become more energized. The proton reacts by raising both of its energy levels, spinning and vibrating, and its position inside of the nucleus. This increases the capacitance between the proton and the neutron so that the frequency or different potential between negative and positive can be maintained. This action decides the positions and energy levels of the electrons orbital shells. By using expansion and contraction the proton helps to maintain a stable atomic structure without the proton itself actually changing in size.

Quantum entanglement by means of the un-energized potential of an Einstein-Rosen bridge, otherwise known as a wormhole, allows the proton to protect the atomic structure. The proton attempting to mirror the perfectly balanced charge of the neutron remains connected to the negative potential of an electron.

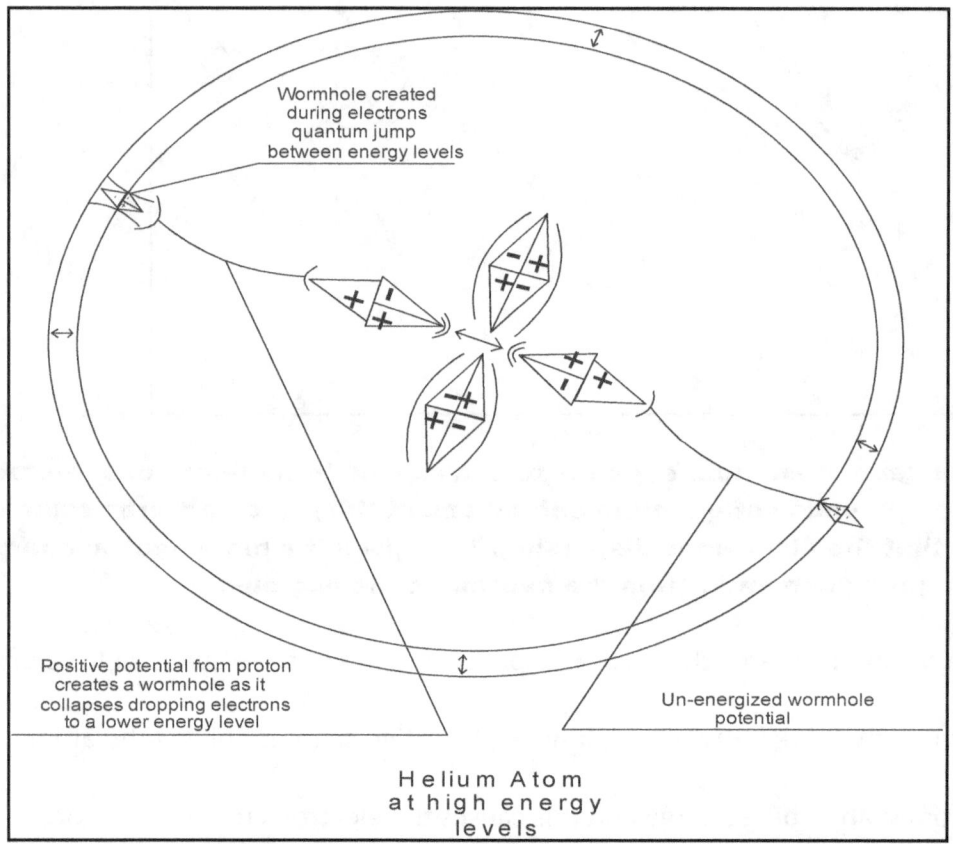

Figure 7- The proton keeps an electron connected through quantum entanglement using the potential of an un-energized wormhole. The proton can push the electron to higher orbital shells when absorbing kinetic energy. As the positive potential collapses back to a lower energy orbital shell a wormhole is created to instantaneously drop the electron to the lower energy orbital shell. This is the proton electron interaction that maintains the atomic structure.

The proton uses wormholes to control the position of the electrons on the outer orbital shells of an atom. The proton will adjust the orbital shell that an electron is using based on its energy level and the energy level of the shell themselves. The act of energizing a wormhole uses some of the excess kinetic energy that has been built up between the proton and the electron. The proton maintains the slightly higher positive potential in the nucleus by dumping this excess kinetic energy and maintains atomic structure and strength as the electron drops to a lower energy orbital shell where it dumps some of the excess kinetic energy as electromagnetic radiation, photons and light. The act of energizing a wormhole uses some of the excess energy of the electron and proton keeping the frequency of the electrical charge between the proton and the electron in a balanced state. This potential between subatomic particles is how an electron manages to stay relative to an atom that it was attached to using quantum entanglement. This entanglement can become a chemical bond between two different atoms or change and be lost once the electron becomes part of a different system. Using an Einstein-Rosen bridge or a wormhole and by matching the frequency of a neutron or that of a perfect balance between positive and negative charges is how the proton uses all of the kinetic energy that an atom absorbs to remain potentially bound to an electron giving an atom the structure and cohesive strength that an atom possess.

Quantum Resonance Frequency is a simple way to describe the fundamental frequency that exists between an electron, a proton, and a neutron as the pure energy that all matter is principally built of creates when the positive and negative charges that are binding this raw energy together form a cohesive bond that is electrically charged giving it the vibration, spin, and kinetic energy that results in mass the basic expression of matter being a separate dimension of electrical charges from the energy that it was created from. The proton uses this cohesive binding force to maintain energy levels and the structural stability of an atom. By combining the positive potential of a wormhole with quantum entanglement the proton keeps the positive to negative charge of an atom stable directing electrons to the energy level or orbital shell that provides the best balance of kinetic energy in the atom. This action creates a basic frequency between the charges of the electron and the proton that will be referred to as Quantum Resonance Frequency.

1.3 The Neutron-

The neutron is the most balanced expression of the energy that became bound by the attraction of the positive and negative aspects that created it. The neutron possesses a neutral charge. This means that a neutron possess both a positive and a negative charge perfectly balanced so that both charges cancel each other out. The neutron uses this balance of both positive and negative energy to act as a type of grounding anchor for an atom. The neutron is not a path of ground like in an electrical circuit, it is more like the instructions that a proton uses to control the energy level and special position of the electron necessary to keep an atoms electrical energy balanced giving the atomic structure the strength that it possesses. The neutron is capable of exhibiting both positive and negative charges equally and in doing so adjusts both its energy level and position in the nucleus in order to adjust to the kinetic energy of the universe around us. This natural reaction to energy levels in the neutrons perfectly balanced state allows the proton to use the neutron as a model adjusting its own position and that of the electrons using quantum entanglement and the potential that connects the proton to the electron with an un-energized wormhole. The neutron uses a balance of positive and negative charges to adjusts its own internal energy levels, vibration and spin, and its position in the nucleus of an atom giving the proton a working model of energy levels and position that allows the atomic structure to have the strength and durability that it possesses and remain a fundamental piece of matter when subjected to the kinetic energy of the electrically charged dimension that we as humans exist in as chemical combinations of matter.

The neutron is not a path to ground for an electrical current but it does use its position inside of the nucleus of an atom in relation to that of the proton to accomplish a similar action. The neutron and the protons position inside of a nucleus can increase electrical conductivity without increasing heat and resistance associated with electrical current. Not every element would be capable of expressing this interaction of the particles position in a nucleus with electrical conductivity but this arrangement can explain the super conductivity of liquid helium. With just two neutrons and two protons at near to absolute zero there is as close to no space between the neutrons and protons as can exist. This allows the two protons to move electrons to higher energy shells using less of the kinetic energy of vibrating and spinning that is being input from an outside source. This ability to use a higher energy orbital shell at next to absolute zero is what gives liquid helium its super conductive like properties. This property of super conductivity is directly related to the position of the neutron and proton in the nucleus giving the atom its atomic structure with their positions being part of the manner that these particles use to keep the positive and negative aspects of the electrical charge inside all atoms balanced.

The neutrons simple balance of positive and negative charges inside of the nucleus of an atom gives it the ability to adjust both its energy levels and position in relation to the kinetic energy of the electrical charge of the universe around it. This balanced charge gives the proton a model to use as it adjusts the position and energy levels of the electrons in the orbital shells of an atom. The very structure of all matter is easily dictated by the neutrons ability to remain a neutral electrical charge. Even the neutrons role in the electrically charged dimension that we exist in appears simple its fundamentally responsible for the atomic structure and therefore the shape of matter that we as humans perceive as solid substance from our point of view inside of this balanced electrically charged system.

1.4 The Electron-

The electron is one of the busiest subatomic particles that we know of and without it our world and all that we know would be completely different. The electron is the negatively charged particle that is working in concert with the proton to form the powerful connective forces of the atomic structure. It is this strong connective force between an electron and a proton that gives atoms both their structure and the electrical charge that we as human beings perceive as solid matter. The electron is orbiting in the outer layers of all atoms moving, spinning, vibrating, and changing distance from the proton it is bound to through quantum entanglement ceaselessly adjusting its energy levels to balance out all of the kinetic energy that all matter in the universe is subjected to. The electron was the first subatomic particle that humans discovered and the reason we noticed it was that we were trying to leave the world of magic and superstition and move into a world of science and knowledge. One of the things that the human race noticed about matter when we abandoned alchemy in favor of chemistry was the strong electrical bond of compound structures like wood, dirt, and rock because of how atoms share electrons to form incredibly strong connections between atoms. These incredibly strong bonds atoms have when they are sharing electrons create a unique frequency and this frequency that pure energy is given with this electrical bond becomes the matter that we see and interact with every day.

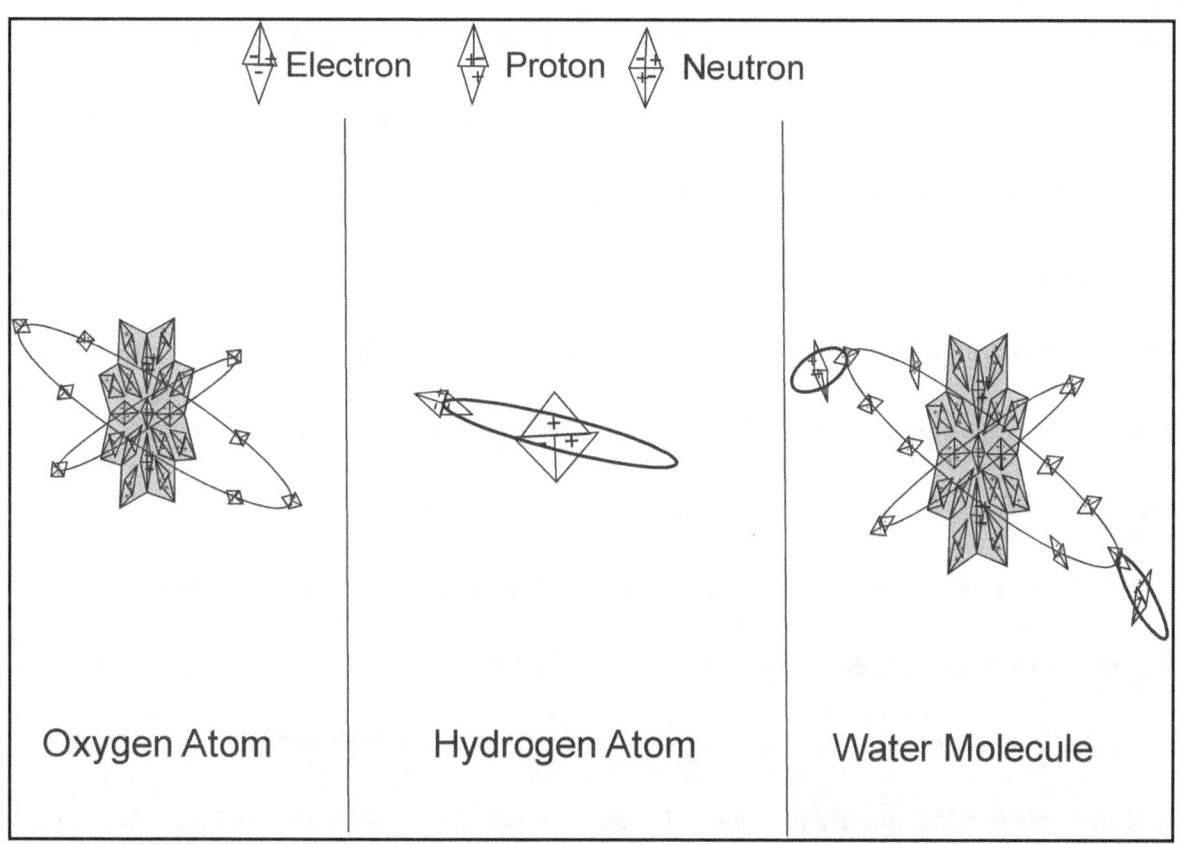

Figure 8- The combination of two hydrogen atoms to a single oxygen atom becomes a water molecule.

The electrical charge that an electron exhibits as it is balancing the positive and negative forces between the proton and the orbital shell the electron is orbiting in creates the electrical field of force that gives matter the solid experience that we associate with the universe around us. This solid field of force is constantly absorbing energy from the universe around the atom. As the electrons and this field of force absorb this energy some of the energy is reflected back out. This is absorption and it is where all colors and texture come from. Without electrons we would have no depth perspective to associate with light and we would not have magnetism that is a further extension of this same field of force protecting the atom. Electrons are the source of electricity and radio frequencies that the human races modern technology is built around and without it our planet could not support the billions of human beings that call earth home. All of this and so much more make the electron the busiest and one of the most vital subatomic particles that we are just beginning to understand from our unique point of view inside of this electrically charged universe.

All negative and positively charged particles have a strong attraction to each other that grows stronger and stronger the closer the charges are from each other. The electron is a combination of a negative charge and a positive charge of pure energy that has the properties that we refer to as mass because of the interaction that takes place between the two charges. The electron is a form of pure energy that has more of a negative charge than it does a positive charge. It is the interaction of these two unique charges that gives the electron and all matter the vibration and electrical quality that we refer to as mass. If the electron was just a negative energy bound to itself it would not constantly respond to the kinetic energy of the universe around us. It is the positive and negative forces constantly interacting with each other that is expressed as the kinetic energy all matter exhibits signs of. The electron is pure energy that is given the properties that we call mass because the positive and negative charges that combine to create one produces and electrical charge that has the dimensional qualities that create a strong electrical field of force around all atoms producing the solid experience that we as human beings perceive when interacting with all forms of matter.

The electron is found on the outer layers of an atom orbiting the nucleus and the positive charge of a proton. There are several energy shells or orbital shells based on energy levels that an electron can use as an orbit depending on the type of atom and the base energy level associated with mass that all atoms are in possession of. Electrons move up and down constantly from one orbital shell to another in and out while orbiting the nucleus. This movement is how electrons use and disperse some of the kinetic energy that the universe is constantly subjecting all matter and atoms to at any given moment. The universe is moving there is a constant amount of energy that comes from the constant movement of all matter. Just the spin of gravity that our planet has moving around the sun creates kinetic energy. This energy is everywhere in the universe it is how kinetic energy is created without matter losing mass creating energy in decay. The field of force that an electron creates becomes a semi solid barrier that is both protecting the nucleus and the balanced positive and negative charges of an atom but it is also strengthening the electrical bond that compounds have when they are joined by sharing electrons. The electrons movement and electrical charge give almost all of the matter that we know of a very strong electrical charge on the surface of all atoms. This means that while all atoms get their structural atomic strength from the powerful connective forces of positive and negative, electrons on the surface of all atoms creates the electrical charge that gives atoms the feeling of solidness that we perceive. All solids, liquids, gases, and plasmas present us with a solid feeling that we can interact with when we come into contact with these substances. Water even though it is a liquid can present a very solid surface if impacted hard enough. Mankind's ability to fly comes from the lift an aircrafts wings receive when moving fast enough through the air. This is actually a solid experience from a medium that we can move through fairly easily. The solid experience that all atoms present us with comes from the similar electrical charge on the surface of all atoms. All atoms possess this exact same electrical charge making it very difficult for atoms nucleuses to actually ever collide. This becomes very important when creating anti-matter. Electrons on the outer shells of atoms provide the very reality that we live in with the solid perspective that we do not even notice unless pointed out because of our unique perspective inside of this electrically charged system that we call our universe.

The electron or combinations of electrons are what form the bonds of chemical compounds turning simple atoms like hydrogen and oxygen into compounds like water or H_2O. Whenever two atoms combine to form compounds it is the electrons orbiting in the outer shells that form this bond. An electrons job in the atomic structure is to absorb the kinetic energy that an atom is exposed to and then to disperse the excess energy. Electrons can disperse energy in many different ways when an atom becomes energized by something more than just the natural kinetic energy that all matter is subjected to. Electrons can create heat, light, colors, and electricity as they orbit the nucleus protecting it from absorbing too much energy. One of the most common ways that electrons absorb and disperse extra energy is with a chemical reaction. Each and every single atom has the ability to absorb and disperse as much energy as the mass associated with the atom. When an atom absorbs more than this amount of energy it can go through a chemical change and join with other atoms. Since these atoms possess the ability to absorb energy equal to their masses combining in ever larger compounds giving these atoms more combined mass to absorb energy than an individual atom possess on its own. To do this atoms will lose electrons shedding some of the energy that it has absorbed with these electrons and then it will join with other atoms by sharing electrons. As this is taking place there is a base frequency inside of the nucleus that the proton is using to maintain the balance between negative and positive charges. During chemical reactions the proton uses the un-energized potential of a wormhole to find the best balancing frequency within the small area of effect that the atom has. All of this excess energy is then transformed between the two atoms and both atoms use the positive potentials to place and hold the electron between the two atoms. Now both atoms are sharing one of more electrons through the potential of the proton electron interaction. This compound now has a different electrical frequency of charges in their combined nucleuses. This new frequency adjusts the atomic structure of both elements in this new compound. This new structure or frequency of the fundamental electric charge that binds the atom together is how two gases like hydrogen and oxygen can become a liquid by the simple sharing of electrons in the orbital shell layers of this pure energy given the perspective that we call mass and matter.

Electrons orbiting the nucleus of an atom create a very strong field of force that absorbs kinetic energy the atom is subjected to and produces the absorption that gives matter the colors that we see when it is energized by photons or the electromagnetic wave spectrum that we refer to as visible light. Light is almost always associated with photons but light comes with two separate distinct aspects. One is the energized photon package that exhibits qualities that we associate with matter having mass and physical characteristics although the photon also possesses some wavelike qualities. The second aspect of light is the electromagnetic wave that gives us as the observer of lights interactions with matter the colors and depths that we are witnessing. The electrons ability to move from shell to shell on an atom based on its energy level is a very important part of light and color as we perceive it. When an atom is exposed to energy in the case of reflected colors this is photonic energy the electrons absorb the energy of the photons as they are energizing the atom. Since the electrons are only absorbing some of the energy of the light that is reflected off of the atoms surface the result is the color that most solid surfaces have. The light reflecting off of a surface maintains the same quality of

electromagnetic radiation as the source of the light.

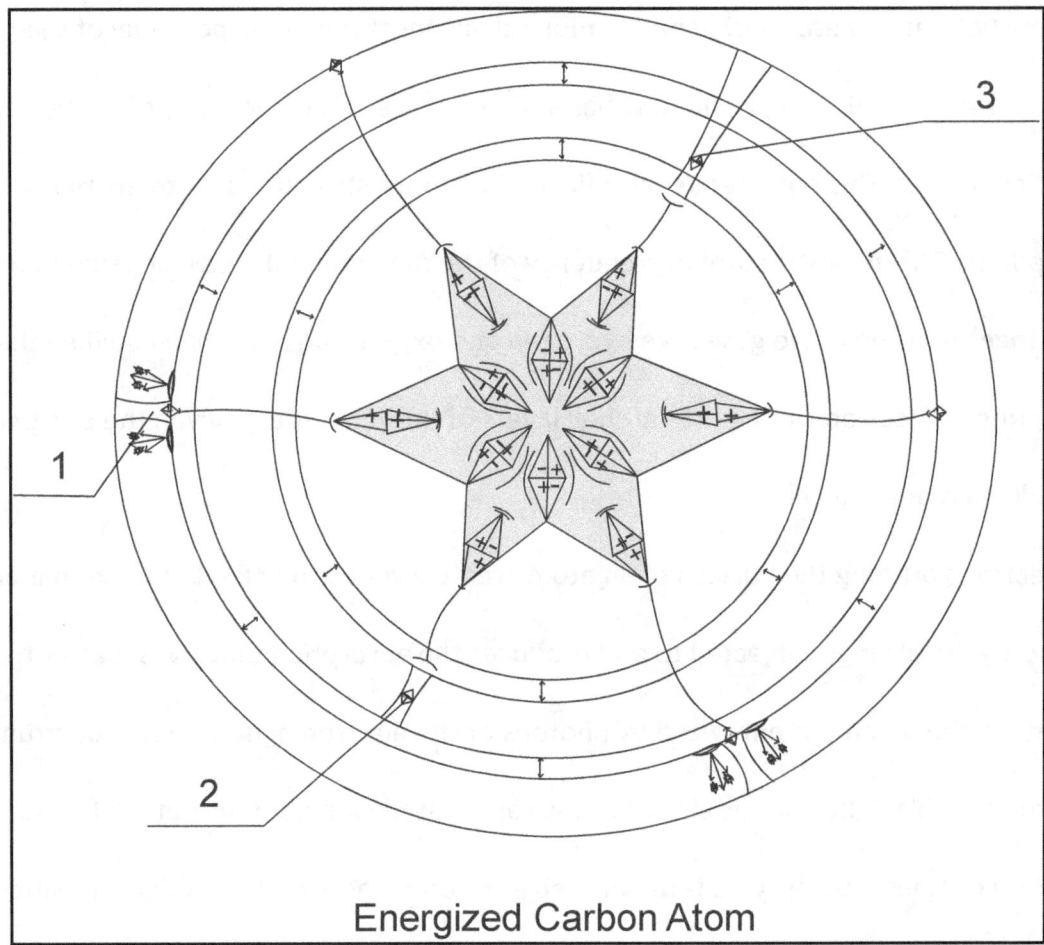

Energized Carbon Atom

Figure 9- Energized Carbon Atom... the protons inside of an atom use quantum entanglement through the connective potential of a wormhole to adjust the position of the electrons in the orbital shells of the atom. The proton moves the electron out to a higher energy level orbital shell as it absorbs kinetic energy. The electron can only stay at the higher energy level orbital shells as long as there are electrons in the orbital shells at the lower energy levels. The Proton will drop the potential that it is holding with the electron to a lower energy level orbital shell to maintain atomic stability when more electrons are moved outwards due to kinetic energy. As the potential between the proton and the electron drops to a lower energy level orbital shell a wormhole opens that the electron uses to jump to the lower energy level orbital shell. All of the excess kinetic energy is given up in the form of light as the electron lands on the lower energy level orbital shell. 1- An electron landing on a lower energy level orbital shell through a wormhole creating light. 2- An electron jumping or dropping down in a wormhole. 3- An electron jumping from the highest energy level orbital shell to the lowest to replace missing electrons.

The surface being energized by the light is absorbing some of the energy of the light and reflecting back the color that is associated with both the energy of the light and the resistance of the surface absorbing the light. This gives most solid matter the colors and textures that we see on any surface. Energized electrons can also produce light of any wavelength or color because of the electrons natural functions on the orbital shells of atoms. The electron is the negative charge that an atom uses to maintain the atomic structure so the electron is connected to the positive potential of a proton. When atoms are in a highly energized state where the atom is producing light it is the electron moving from one orbital shell to another based on the energy level of the electron that is creating both the electromagnetic radiation and the photons associated with light. The electron uses all of the excess energy that it has to move back and forth from one orbital shell to another using quantum entanglement with the proton it is associated with and wormholes to jump back and forth in orbital shells. This is called a quantum jump. The proton uses the excess energy that the atom is being subjected to in order to keep the electron in the position that has the greatest stability to the atomic structure. The electron gains energy at a lower orbital shell and then is moved outwards to the orbit that is associated with the kinetic energy being applied to the electron. The atom still has to maintain the structural integrity that comes from the position of the electrons at the lower stable energy level shells. This causes the potential held by the proton to drop down to the lower energy level orbital shell to maintain the structural integrity of the atom. The electron will jump into a wormhole created by the proton dropping down to the lower energy level orbital shell and end up on the lower energy level orbital shell. Since the electron is still highly energized it will create light both electromagnetic and photonic at this orbit losing even more energy and dropping down to an even lower kinetic state. The combination of energy levels is what gives light the specific spectrum or color that it is associated with. A very simple way to view this is by viewing the orbital shells of an atom like guitar strings. Instead of producing sound the orbital shells of atoms are producing light depending on the orbital shell or combination of orbital shells the electron uses to disperse the energy that an atom is being subjected to. A single string or orbital shell of an atom will produce a blue or a red colored light. A combination of both strings or orbital shells will produce a violet colored light. The energy level of the electron as it jumps to an orbital shell will also affect the electromagnetic radiation and the color or wavelength of light. This can viewed as how hard a guitar string gets plucked. The electrons field of force that it creates surrounding an atom is directly related to the quality of the colors and the light that we see in the world around us.

Most if not all of the energy that an atom is subjected to has a negative quality thereby producing a universal electrically charged dimension that is reflected in the nature of most kinetic energy. The outer orbital shell of all atoms is made up of electrons. These electrons produce strong fields of force that surround all atoms. These strong fields of force have a negative quality due to the electrical charge of the electron creating these fields of force. All atoms and matter in the universe is constantly moving bumping and colliding over and over and over. The collisions of all of this matter and all of the interactions that take place at a distance are where all of the kinetic energy that all atoms are subjected to comes from. Since all atoms possess a very strong negative field of force it is these strong fields surrounding atoms that are colliding and creating this predominately negatively charged kinetic energy. This is the major reason why the proton is 1800 times larger than the electron. The atom maintains its atomic structure due to the interactions and strong cohesive force of positive and negative charge balanced out and bound together. Almost all kinetic energy that any matter is subjected to is negative the proton has to maintain the positive charge that binds the atom together. This leaves the electrons negative charged mass at somewhere around ten percent of that of the proton when it is fully energized. The kinetic energy of the universe around us produces enough of a negative charge that the atomic structure of an atom maintains its internal strength and structure by maintaining around ninety percent of the electrical charge of an atom as a positive charge bound in a proton and the rest of the electrical charge of the atom as a negative bound in the electron plus all of the kinetic energy all around the atom.

The electron performs a function called a quantum jump as it moves from energy level to energy level and orbital shell to orbital shell. The electron moves in a fashion that can be described as faster than the speed of light as it is jumping from one orbital shell to another orbital shell. We know that due to the laws of physics that nothing can travel faster than the speed of light. We know based on the laws of sanity that electrons do not go into an alternate universe, hang out drinking coffee and then magically return at some distant point in a finite amount of time that is shorter than the amount of time that it would take for light to travel the same distance in question. Electrons are using an Einstein-Rosen bridge based on the positive potential of a proton balancing out the potential charge of an atom based on the distance of the electron from the proton. The protons in the nucleus of an atom maintain a connection to the electrons in the orbital shells by quantum entanglement that comes from the balanced potential of the positive and negative charges connected through an un-energized wormhole. When an atom is absorbing the negative kinetic energy that produces the electromagnetic energy of light, the proton has the ability to absorb some of this energy with the spin, vibration, and its position in the nucleus of the atom. Once the proton has become energized the electron is forced outwards to the point that best matches the balanced frequency of the negative to positive ratio of the atom. The potential of the strong connective force causes a wormhole to become energized and the electron jumps to the appropriate orbital shell that matches the new frequency or energy level. This causes the electron to lose some of the excess kinetic energy and once again the proton adjusts its position and energy levels. The electron performs another quantum jump landing on a lower orbital shell and giving up some more of the excess kinetic energy in the form of electromagnetic radiation and the creation of photons and light. This process is continuous in sources of light as the atoms natural strength and structural integrity allow the electron to move throughout the orbital shells surrounding an atom absorbing and dispersing kinetic energy in a concerted effort with protons to keep a perfect harmonic balance between the positive and negative electrical charges that are the cohesive forces binding all atoms and matter together with the properties that we refer to as mass.

Electricity, radio waves, and radio frequencies are produced by electrons as they move throughout our universe performing their basic task of protecting the atomic structure of atoms. Electricity is the movement of electrons from one valence layer to another in a continuous path as the electrons are attempting to disperse the kinetic energy that was originally applied to give them the electric momentum to start with. It is the electrons natural disposition to absorb the negative kinetic energy that all matter is subjected to that gives us the electricity that we use every day. Electricity was first noticed as lightning bolts crashing to the earth back when humans still thought that lightning bolts meant that the gods were angry.

Electricity was first manipulated with static discharges that humans noticed when rubbing amber and fur together. The fascination at the static produced and how it resembled lightning bolts caused the human race to invent, design, and promote ways to produce electricity. Now today we have technology that has allowed the human race to communicate vast distances and even leave our tiny little planet to explore our solar system and begin to view the greater expanse that we call our universe. All of this technology that we have created and mastered stems from the simple electron and one of its basic functions where its movement, energy levels, and position are entirely determined by the positive and negative ratio of the internal frequency of the nucleus of an atom. Even when electrons leave the valence layer of an atom either as radio waves or electricity it is because the overall negative charge or kinetic energy that an atom has been subjected to has reached the point that ejecting electrons becomes the best way to maintain the atoms basic structure. We have determined which elements are the best conductors and insulators to promote the use of electricity but the properties that give us electricity and insulation are based on the atomic structure and the placement of electrons in orbital shells. The metals that produce the best electric currents have extra electrons on the outer orbital shells. The compounds that produce the best insulators have a balanced outer orbital that does not readily accept extra electrons. When the atom is producing electricity it is just maintaining the basic atomic structure where the kinetic energy that the atom is absorbing is predominately negative and in order to maintain the balance of positive to negative charges the electrons leave or flow across the exterior or valence orbital shells. Radio waves are simply an excess number of electrons without a path on the conductive surface that will allow for atomic stability these electrons are merely ejected off of the surface of the atoms subjected to this excess kinetic energy. Humans have learned to control this function of the atom because of the precise mathematical repetition of the basic atomic structure. We have manipulated this form of atomic structural security for communication purposes to the point of almost every human being on the planet being can communicate with each other at any time they choose. All of this is because the positive charge of the nucleus maintains a perfect symmetrical balance with the negative charge of the electrons in the atoms orbital shells constantly absorbing the kinetic energy of the universe around us. We has human beings exist in an electrically charged dimensional state of pure energy that has been given form and mass because of the vibrational force exerted from the joining of a positive and negative charge bound together to create matter as we know it. Electricity and radio waves are simple results of the nature of a combined positive and negative charge given mass by the strength of their connective forces.

Electrons have so many roles throughout the universe that listing each and every one easily fills up entire books on chemistry, physics, electronics, and cosmology. Electrons are basically the first thing about all matter that we will experience and interact with due to the structure of an atom. All matter is subjected to kinetic energy throughout our universe because of the internal energy of all matter constantly being in motion. Where atoms are concerned this leaves us with a very strong negative charge on the exterior of all atoms due to the strong field of force that electrons create in the orbital shells of atoms. As atoms bounce around, collide, vibrate, and interact with each other this principally negative charge on the outside of all matter comprised of atoms is maintained, absorbed, and dispersed by the electrons that surround the nucleus of an atom. The cohesive strength of a positive and negative charge bound together is the very basis of the atomic structure of all elemental matter that we know of. The volume of negative kinetic energy that all atoms are continuously subjected to can be seen as the size of the electron compared to the size of the proton. At 1800 times smaller than the positively charged proton the electron is still capable of maintaining the negative charge necessary to maintain the atomic structure because of the constant kinetic energy that it is subjected to. Electrons are the source of the strong field that surrounds matter that we interact with and the reasons why we perceive the structured energy of atomic mass as being solid. As long as there are science books the electron will hold a key role in any scientific study of our universe because of the simple role it plays in maintaining the atomic structure of all elemental matter that we interact with every day.

1.5 Chemical bonds and the dimensional aspects of the nucleus-

Every atom has its own unique frequency and signature that changes based on the energy level of the compounds involved. The shape of an atoms nucleus is obviously based on the number of protons and neutrons that form the center of an atom. The position of the protons and neutrons and the movements relative to the energy levels that an atom is being subjected to gives the nucleus its dimensional properties or shape. This can be viewed entirely as quantum frequencies or quantum properties based on the number of protons, neutrons, electrons and the energy levels and positions of each in relation to the bound potential of a positive and negative charge giving the atom its basic atomic structure. In essence the basic shape of the nucleus is determined by its size and the kinetic energy that the atom is exposed to this shape changes based on the movement or energy levels of its constituent parts; the proton, electron, and neutron. Each and every shape or chemical bond is a different quantum frequency giving all matter the infinite number of shapes, textures, colors, and strengths that we refer to as chemical compounds or just the everyday world around us. As the simple frequencies in the nucleus change so does the very fabric of an atomic structure of matter. Quantum Resonance Frequency is based on the quantum frequency produced by the varying dimensional states of the nucleus of atoms whether positioned individually or part of a complex compound that gets its shape, color, and texture from all of the different quantum frequencies combining to produce the matter that we see every day.

The generally held belief about the shape of an atoms nucleus comes from the number of protons and neutrons and what we have witnessed in an electron microscope. The exact position of each proton and neutron in a nucleus can only be generalized due to the fluid like movement of the nucleus. Instead of viewing protons and neutrons as solid matter both should be viewed as something similar to plasma. Neither protons or neutrons possess a true shape of their own. Instead their relative dimensions will be based on their own internal reaction to all of the other protons and neutrons in the nucleus and by the energy levels and interactions with the electron as the basic atomic structure keeps the cohesive balance of positive and negative charges. This fluid like complex structure will possess a limitless number of possible dimensional shapes giving all known matter the complexity of the diverse shapes, colors, and textures that we see and interact with on a daily basis.

1.6 Mass-

Mass is the existence of matter in our electrically charged universe. Mass is the determination of how much of a substance actually exists. At very large chemical combinations gravity, weight, and volume can be used to determine the general size of most forms of matter. When we are dealing with substances as small as atoms we use mass and atomic weight ratios. Mass for subatomic particles is the vibrational or electric quality that pure energy has when a single unit or small quanta of a positive and negative charge are bound together. All particles possess an electrical charge. In the case of protons this charge is a positive charge. For neutrons the charge is a neutral charge. Electrons have an overall electrical charge that is negative. All three subatomic particles possess and electrical quality that we refer to ass mass. The large amounts of pure energy that become bound in single quantities of electrical charges creates a vibration or ripple throughout space that gives pure energy the proportions that we define as mass. This becomes part of the strong cohesive force that is the basic atomic structure. All protons, neutrons, and electrons possess similar electrical qualities that we refer to as mass. This creates a slight repelling force for subatomic particles. This electrical quality is combined with the electrical charge. The positive and negative charges of subatomic particles have a very strong attractive force. The combination of electrical quality and electrical charge is the basis of the atomic structure. Electrical quality or the minute vibration that pure energy has when a single quantum of positive and negative charges are bound together in subatomic particles is where the mass that we use to define our universe comes from.

1.7 Quantum resonance Frequency-

Quantum Resonance Frequency is a theory based on the atomic structure of all matter where the pure energy that all matter is made of is given mass by the cohesive bond of a positive and negative charge bound together; meaning that all atoms, molecules, and chemical compounds possess an exacting frequency that is derived from the energy levels of the basic interpretations of protons, neutron, and electrons as the dimensional properties of volume, shape, size, and strength change in exacting proportions to the quality of the atomic structure of the nucleus the constituent particles and the strong field of force created by the electrons as they absorb and disperse kinetic energy. The shape of the nucleus is directly proportional to the number of protons and neutrons. The protons and neutrons inside of all atoms have the ability to change their energy levels and position inside of the nucleus to correspond with the energy level of the electrons as they are dispersing the kinetic energy that they are subjected to. The neutrons role as a balanced positive and negative charge allows the proton to hold a positive potential and position in relation to the electrons in the orbital shells of the atom. The proton uses quantum entanglement through an un-energized wormhole to maintain an atoms balanced positive and negative charges by controlling the positions of the electrons in the orbital shells of all atoms. The electron is the workhorse of human technology because of its position in the orbital shells of atoms and the natural ability it has to produce light or become electricity and radio waves as it is constantly producing the strong field of force that we perceive as the solid reality around us. All atoms, elements, molecules, and chemical combinations thereof produce a quantum frequency that is the exact representation of all of the physical characteristics that we define as mass and the properties of matter due to the constant flux of the pure energy bound by the cohesive strength of a positive and negative charge. This pure energy that constitutes all known matter displays unique properties depending on the simplicity or complexity of all of the constituent molecules, atoms, particles, or just the sheer number of individual electric charges that work in unison to create the electrically charged dimension that we call our universe. The simplest way to describe the complexity of all the different quantum dimensions capable of being produced by the cohesive bond of pure energy bound by its positive and negative aspects is Quantum Resonance Frequency.

Viewed in the simplest terms an atom is pure energy bound together by the strong bond of a positive and negative charge balanced by the force exerted due to the similar electrical qualities of mass. What we perceive as the mass matter has is a vibrational quality that can be viewed as kinetic energy. Pure energy bound together by the positive and negative charges of its characters produce a constant flux between these two components. Since this pure energy is experiencing no resistance there is no loss of energy just a constant rhythmic flux that gives the combination of charges the properties that we refer to as mass. The position of the positive and negative particles that produce this mass creates a strong electrical charge. All matter that we are capable of interacting with possesses this same electrical quality. We perceive matter as being solid because the negative charge created by electrons on the surface of all atoms is exactly the same for everything we know of in our electrical charged dimension. This means that the dimension that we live in and all of reality that we can view and interact with possesses the same exact electrical charge. The simple rule of similar charges repelling each other creates the strong field of force that we perceive as being solid. This is also why particles that do not possess the same electrical quantity or charge pass through everything that we call solid matter. Most of these particles can be considered out of phase with the electrically charged dimension that we live in. All matter is pure energy that has the properties that we define as mass and matter due to the quality of the electrical charges and dimensional aspects this energy possess through complex combinations of vibrational qualities as more and more mass and electrical charges build up to ever increasing chemical combinations.

The nucleus of an atom is where different combinations of neutrons and protons maintain the balance between the positive and negative properties of the atom. The neutron is a neutrally charged particle but it is still a charged particle. This means that the neutron possesses both a positive and negative charges bound equally producing a balanced or neutral charge. This neutrally charged particle allows the atomic structure to increase in size from a single proton and a single electron. The reason why atomic mass increases proportionally to the number of neutrons in an atom is because the kinetic energy of the atomic structure could not be maintained with just protons. The flux created by the positive and negative aspects of this pure energy would burn up the nucleus of an atom without neutrons to help maintain the balance between protons and electrons in elements larger than hydrogen. Similar to an insulator the neutron maintains the balanced space between protons inside of the nucleus of an atom. The size and shape of an atoms nucleus is determined by the level of positive and negative charges that are being balanced out to maintain the atomic structure.

The proton in an atoms nucleus is the primary source of the positive charge inside of the atomic structure. A proton is a positively charged particle this means that a proton has both negative and positive aspects of the electrical charge that gives it mass. The proton has more of a positive electrical charge than a negative charge. This gives the energy that produced the proton properties that we associate with having mass. The proton uses this vibrational quantum flux to maintain a perfect symmetry with an equally proportioned negative charge. This results in quantum entanglement through the use of un-energized wormholes that become energized as the electron performs quantum jumps associated with light while dispersing excess kinetic energy the atom is subjected to. The proton uses its position and energy levels to maintain the distance and positive potential in relation to the electron and the field of force that the electron creates in the orbital shells. The size of an atom can increase if excess energy is applied to it. This usually produces heat and low level photonic interaction we refer to as expansion and contraction. Using quantum entanglement the proton actively moves the electron to the best orbital shell position necessary to maintain the balanced symmetry of positive to negative that gives an atom the atomic structure that it possesses. The dimensional shape of the nucleus of an atom and the positions of electrons in complex chemical combinations give the atom the very aspects that produce size, shape, and volume is directly related to the protons position and energy level inside of an atom.

The electron is the most easily recognized force of any particle that makes up matter other than a photon. The electron is a negatively charged particle. This means that the electron has both positive and negative aspects but has a greater negative potential to the electrical charge that gives this particle both its mass and wavelike properties. Our society has become dependent on the electricity and technology that is produced by manipulating the natural characteristics of the electron. The electron orbits atoms in various orbital shells maintaining the negative aspects of the atoms structure. The electron also creates the strong field of force that is associated with all electrical charges. Similar charges repel each other and dissimilar charges attract each other. The surface of all atoms in the electrically charged dimension that we perceive as solid matter comes from the electrons orbiting atoms. The very sense of solidity that our universe presents us with is derived from this basic law of physics. Our ability to manipulate the electron comes from the electrons natural ability to absorb and disperse excess kinetic energy. Since most of the kinetic energy that all matter is subjected to is of a negative quality derived from the collisions and interactions of the exterior of atoms the electron is 1800 times smaller than the proton. The vast difference is size is directly related to the proton holding the positive charge and the surface of all matter in our electrically charged dimension having aspects of the negative charge. The electrons size keeps the negative charge of an atom from burning up or coking the basic atomic structure. The electrons position in the atom and its natural disposition to disperse excess kinetic energy has made this subatomic particle the workhorse for the human race and our energy dependent society.

Quantum Resonance Frequency is a simple way to view the complexity of the atomic structure. The cohesive strength of an atom and all of its constituent particles is directly derived from the strong bond that pure energy has in relation to its own positive and negative aspects. The essence of matter itself that we define as mass is nothing more than the vibrational quality of the pure energy that is bound by its very own potential differences. This requires the human race to use mass as a measure of quantity when attempting to explain or interpret the reactions and interactions of this type of pure energy. If all energy throughout the universe were viewed as one infinitely long wave then the atom would be a single spike in the wave containing both a positive and negative charge. Space becomes the intervals of time between these spikes of energy leaving the quantity of all matter as the true frequency since all matter is said to have been created at the instance of the big bang. Einstein's equation easily expresses this very notion. E(energy)=M(mass)*C2(the speed of light multiplied times itself). The relationship between energy and matter has already been established by some of the most brilliant minds the human race has had. Understanding that all matter is just the vibrational qualities of pure energy when it becomes bound by its own potential becomes a small step in our understanding once the basic principals have been established. Quantum Resonance Frequency is a theory that is based on the attempt to understand the basic nature of our universe by viewing all matter as raw energy given mass and dimensional aspects because of the strong cohesive force of a positive electrical charge bound to a negative electrical charge.

References

ATKINSON, N. (2011, January 10). *Fermi Telescope Catches Thunderstorms Hurling Antimatter into Space*. Retrieved January 2011, from universetoday.com: http://www.universetoday.com/82369/fermi-telescope-catches-thunderstorms-hurling-antimatter-into-space/

Barry, P. (2010, February 2). *Firefly Mission to Study Terrestrial Gamma-ray Flashes.* Retrieved March 2010, from PHYSORG.com: WWW.physorg.com/news184313955.html

Barry, P. (2010, February 2). *Firefly Mission to Study Terrestrial Gamma-ray Flashes*. Retrieved March 2011, from phys.org: http://phys.org/news/2010-02-firefly-mission-terrestrial-gamma-ray.html

Bonnell, J. (2002). *A Bad Day in the Milky Way*. Retrieved March 2011, from pbs.org: http://www.pbs.org/wgbh/nova/gamma/milkyway.html

Carl Zorn, D. S. (n.d.). *What keeps the electrons revolving around the nucleus of an atom?* Retrieved May 26, 2011, from Education, Thomas Jefferson National Accelerator Facility - Office of Science: http://education.jlab.org/qa/atomicstructure_08.html

Cohen, D. (2007-2011). The Universe. *The Universe Anchient Mysteries Solved-series* . USA: Louis C. Tarantino, Douglas Cohen.

Copper Development Association. (2009). *Copper Production from Ore to finished Product*. Retrieved February 22, 2009, from Copper.org: http://www.copper.org/education/production.html

Cowen, R. (2011, January 10). *Today's weather: thunder and antimatter beams* . Retrieved November 7, 2011, from sciencenews.org: http://www.sciencenews.org/view/generic/id/68592/title/Todays_weather_thunder_and_antimatter_beams

Encyclopædia Britannica. (2011). *plasma.* Retrieved November 20, 2011, from Encyclopædia Britannica: http://www.britannica.com/EBchecked/topic/463509/plasma

Encyclopedia Britanica. (2014, 1 29). *Hydrogen peroxide*. Retrieved August 1, 2014, from Encyclopedia Britanica: http://www.britannica.com/EBchecked/topic/278760/hydrogen-peroxide

Encyclopedia Britanica Inc. (2011, March). *Valence Electron*. Retrieved March 2011, from Encyclopedia Britanica: http://www.britannica.com/science/valence-electron

Morris, N. M. (1991). *Mastering Electrical Engineering.* MACMILLAN EDUCATION LTD.

NASA. (2011, November 1). *FERMI Gamma Ray Telescope.* Retrieved November 7, 2011, from NASA.gov: http://fermi.gsfc.nasa.gov/

NASA. (2010, January 29). *Firefly Mission to Study Terrestrial Gamma-ray Flashes* . Retrieved Novmeber 7, 2011, from science.nasa.gov: http://science.nasa.gov/science-news/science-at-nasa/2010/29jan_firefly/

NASA. (n.d.). *Gamma-Ray Bursts: Introduction to a Mystery.* Retrieved 2005, from NASA.gov: http://imagine.gsfc.nasa.gov/docs/science/know_l1/bursts.html

NASA. (2011, Janruary 10). *NASA's Fermi Catches Thunderstorms Hurling Antimatter into Space.* Retrieved january 10, 2011, from NASA.gov: www.nasa.gov/mission_pages/GLAST/news/fermi-thunderstorms.html

National Geographic Society. (2011). *Lightning.* Retrieved November 20, 2011, from environment.nationalgeographic.com: http://environment.nationalgeographic.com/environment/natural-disasters/lightning-profile/

NDT. (n.d.). *Gamma Radiation.* Retrieved February 2011, from NDT Resource Center: http://www.ndt-ed.org/EducationResources/CommunityCollege/Radiography/Physics/gamma.htm

Net Industries. (2011). *Acceleration-History, Linear Acceleration, Cirular Acceleration, Force and Acceleration.* Retrieved January 21, 2011, from jrank.org: <a href="http://science.jrank.org/pages/12/Acceleration.html">Acceleration - History, Linear Acceleration, Circular Acceleration, Force And Acceleration</a>

New Scientist Magazine. (2004,2005). *New Scientist* .

Newton, S. I. (2010 (1687)). *The Principia.* snowball publishing.

Pauling, L. (1988). *General Chemistry.* San Francisco: Dover.

Rosenberg, P. (2005). *Basic Electronics.* Indianapolis: Wiley Publishing,Inc.

ScienceDaily LLC . (2009, July 8). *Antimatter Positrons Explain Gamma Ray Mystery In Milky Way Galaxy.* Retrieved 2011, from ScienceDaily: http://www.sciencedaily.com/releases/2009/07/090708201840.htm

Smith, H. R. (2013, September 18). *What Is the Fermi Gamma-ray Space Telescope?* Retrieved December 2011, from NASA.gov: http://www.nasa.gov/audience/forstudents/k-4/stories/what-is-the-fermi-telescope-k4.html#.VXy3d_lViko

Stephens, T. (2005, February 21). *New satellite observations of terrestrial gamma-ray flashes reveal surprising features of mysterious blasts from Earth.* Retrieved November 7, 2011, from http://currents.ucsc.edu/04-05/02-21/flashes.asp: http://currents.ucsc.edu/04-05/02-21/flashes.asp

Steve Gagnon. (2015). *Questions and Answers*. Retrieved August 11, 2015, from Jefferson Labs: http://education.jlab.org/qa/proelesize_01.html

Sutton, C. (2015). *Antimatter/Physics*. Retrieved June 13, 2015, from Encyclopedia Britanica: http://www.britannica.com/science/antimatter

US Department of Energy. (2014). *Hydrogen Production*. Retrieved July 31, 2014, from Department of Energy: http://energy.gov/eere/fuelcells/hydrogen-production

US Department of Energy. (2014). *Increase Your H2IQ*. Retrieved July 31, 2014, from Energy.GOV: http://energy.gov/eere/fuelcells/increase-your-h2iq